What is ecology?

All living things that are known to exist are found on one planet, the Earth. They all share this planet, from bacteria too small to be seen without a microscope to the giant redwood trees and the whales of the oceans.

All the living and non-living things that surround such a plant or animal are called its environment. For example, the environment of a plant includes the soil, the water and foodstuffs in the soil and the air the plant is growing in. Rainfall and temperature may affect the life of the plant, so may other plants that compete for water and food. There may also be animals that eat the plant and some that may help it to reproduce. All these things make up the plant's environment. The science that looks at the ways in which plants and animals affect their environment, and are affected by it, is called "ecology."

URBAN ECOLOGY

Jennifer Cochrane

Series Consultant: John Williams
Series Illustrator: Cecilia Fitzsimons

The Bookwright Press
New York . 1988

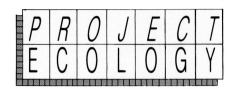

Air Ecology
Animal Ecology
Land Ecology
Plant Ecology
Urban Ecology
Water Ecology

First published in 1987 by
Wayland (Publishers) Ltd
61 Western Road, Hove
East Sussex BN3 1JD, England

© Copyright 1987 Wayland (Publishers) Ltd

First published in the
United States in 1988 by
The Bookwright Press
387 Park Avenue South
New York, NY 10016

ISBN 0–531–18156–1

Library of Congress Catalog Card Number
87–70041

Typeset in the UK by
DP Press Ltd, Sevenoaks, Kent
Printed in Italy by
G. Canale & C.S.p.A., Turin

Cover: bottom *wallflowers, aubretia and white deadnettle growing in a crack in a wall;* left *a swallow feeding its young on a city ledge;* right *trash in a New York City street.*

Frontispiece: *Blackbirds nesting in a pedestrian crossing sign.*

Contents

1. Deserts of concrete

When humans first evolved on Earth, they did not stay in one place. They wandered in groups, sheltering in caves and hunting animals and gathering wild plants. These wandering hunters lived through the last Ice Age, 10,000 years ago. They decorated the walls of their caves with paintings, and made tools, weapons and clothes, but they did not build places to live.

Eventually, these nomadic people settled down and became farmers, living in small communities that grew into towns. The first city appeared some 4–5,000 years ago when enough food could be produced to support specialist workers other than farmers. Early cities were walled for defense, and humans began to make considerable changes to the habitats in which they settled. The wildlife was first reduced, and then almost eliminated, as the towns grew larger and more built-up.

A view over New York at dusk. Cities consume huge quantities of energy. Food and goods have to be transported over long distances into them, and wastes and sewage out of them.

Plants and animals did not disappear from towns. Cattle, sheep and poultry provided milk and meat, while plants were cultivated and grain was stored, which attracted rats and mice. Cats were kept because they hunted the grain-eating rodents. In ancient Egypt cats were worshiped because their hunting protected the stored grain.

As towns and cities grew too large for the surrounding areas to support them, trade with other countries became important. Until the Industrial Revolution only one person in five lived in a town of over 10,000 people. In the past 100 years, though, there has been a great flow of people into "urban settlements." In 1900 only about 14 percent of the world's population lived in cities. By 1950 the world's urban population had more than doubled, and

By the year 2000 it is thought that Mexico City will be the largest city in the world, with over 30 million inhabitants. The prosperous inner-city is already ringed by vast "squatter camps" of self-built dwellings.

by the year 2000 over half the world's people will probably live in urban areas.

Cities have a strong effect on the environment, covering large areas of land, polluting the air and rivers and using up large amounts of energy, fuel and materials just to keep them working. With apartment and office blocks, crowded streets and a polluted atmosphere you would hardly expect to find much wildlife. Yet a city provides a varied environment for its inhabitants, and living things have adapted to urban life.

2. Urban air

Air is not naturally clean since there will always be dust and pollen mixed up in it. Yet other substances from the burning of fuels can make town air very polluted.

Polluted city air contains carbon monoxide, a poisonous gas produced when gasoline is burned; lead may also be present. This metal is often added to gasoline to make engines run better but it is a dangerous substance. In many cities about 90 percent of the lead people breathe comes from vehicle exhausts. Lead can seriously affect people's health and cause damage to the brain, liver and kidneys. Several countries have stopped putting lead in gasoline, and others are phasing out its use.

Other gases from vehicle exhausts include nitrogen oxides and hydrocarbons. These can be dangerous on their own or when they combine to give the poisonous smog containing ozone, which affects cities around the world from Australia to Peru. Such a smog was first noticed in Los Angeles, in 1943. It has now been reduced by fitting antipollution equipment on car and truck exhausts. However, more than 40 percent of nitrogen oxides in the air come from vehicles and can cause health problems for city-dwellers.

It is not only humans who suffer problems from polluted air. The lungs of town birds and mammals are often gray or black because of the dust and smoke they breathe in. Dirt can clog up the pores in the leaves of plants, preventing the necessary evaporation of water from them. Dust and smog can also cut down the light reaching the leaves for photosynthesis.

The ozone produced in certain smogs is thought to be contributing to the death of large areas of forest throughout Europe.

The smog of the kind formed over Los Angeles has extensively damaged forests in the region and posed a serious health hazard.

Activity: How clean is town air?

What you will need

Some white cardboard or posterboard, petroleum jelly (vaseline) and string.

Take the piece of cardboard and draw rows and columns on it, about ½ inch apart, so you have a number of squares. Cover the surface with a thin layer of petroleum jelly and pierce two holes so the string can be threaded through, as in the diagram.

Make several such grids and leave them in different places. One could be left in a busy room, another in a room rarely visited. One could be tied to a tree or lamp post near a road and another fixed on the outside of a window. Try to put them in as many different places as you can think of. Make sure some are facing the wind and others are sheltered by walls.

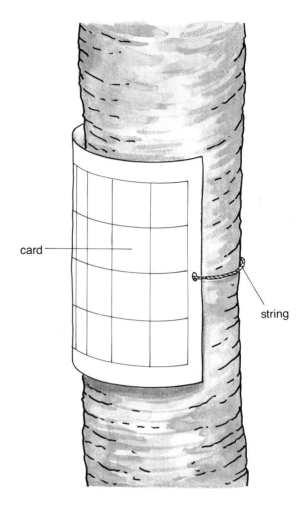

card —

string

Leave the grids for a day and then collect them. Use a magnifying glass to count any specks of dirt on one square of the grid. Show your results in a table and grade the places they were situated according to dirtiness.

place	number of dirt particles per square of the grid
roof	
lamp post	
near road	
playing field	
window (not facing the wind)	
window (facing the wind)	
wall (inside)	
room that is not used	

What did you see?

What did you find on the grids? Which were the "dirtiest" places and which the "cleanest"? What is the reason for this? How can you account for your results in terms of the wind direction and possible sources of pollution?

3. Town climate

For many reasons towns have a climate of their own quite different from the countryside around them. The pollution in the air is one reason, since it reflects up to 20 percent of the Sun's energy falling on it.

A more important factor affecting a town's climate is the stone, concrete and asphalt that make up its surface. In open countryside the Sun's energy is absorbed by the land and much is held there during the day. In towns, the sidewalks and roads reflect more of the Sun's energy than the land surface in the countryside does, and materials like glass and metal reflect even more.

In built-up areas concrete, asphalt and glass reflect much of the Sun's energy falling on them, and rainwater is quickly channeled into sewers. This is the road leading to Sydney Harbour Bridge, in Australia.

Some of the heat is trapped by the polluted air so the warmed air rises and forms a "heat island" over the town. Large towns are, on the average, warmer than the surrounding countryside. In the spring you will find leaves appearing on city trees while in the country they are still in tight bud. Urban areas are also less cold during the winter, and places like parks that are ringed by buildings will escape severe frosts, making it easier for the wildlife to survive.

Within the town the temperature can be quite uneven. Parks, lakes and rivers are cooler areas while walled and paved places can get very hot indeed, giving smaller heat islands within the town. Large towns also suffer less from strong winds than rural areas. However, the tall buildings create many updrafts and breezes, which architects have to take into account when designing buildings. The circling motion of leaves and litter near

Starlings roosting on a city building. These birds flock into cities at night, possibly to take advantage of the higher temperatures in urban areas.

buildings shows the effect buildings have on the wind.

Cities also have slightly more rain than the surrounding countryside. Since most of the rain falls on roads and buildings it quickly evaporates back into the atmosphere. The remaining water passes through drains to streams and rivers with a little absorbed by the soil in town parks and gardens. As a result, towns can be quite dry places despite having slightly more rainfall, and more fog and cloud, than country areas. How these differences affect wildlife is uncertain, but the habit of birds like starlings of flying into towns to roost at night is probably due, in part, to their need for warmth.

4. Water in towns

Towns were originally built in places with a good water supply. In the past water came from wells, streams and rivers in the town. As the towns grew in size it became necessary to purify the water and bring it in from distant areas in pipes.

Rivers are sources of water for our homes and factories. They carry surplus water from the land, preventing flooding, and they provide places for recreation and wildlife. However, rivers are also used for the disposal of liquid waste, and a city produces a lot of this.

The discharges from sewage plants and waste from paper and food factories all find their way into rivers. If the amount of this waste matter is small compared to the amount of water in the river, it can be broken down by bacteria and fungi. These will use up oxygen

The diagram below shows how the absence of some species from water can indicate the level of pollution, or lack of oxygen.

dissolved in the water to purify the river again. The oxygen is replaced since it dissolves into the water from the air, and is given off by green plants in sunlight. If the pollution is heavy, though, oxygen will be removed faster than it can be replaced.

The animals in the river all depend on the dissolved oxygen for breathing and they will quickly die if there is not enough.

A river can be damaged by other pollutants, including fertilizers and pesticides washed from farm land. In late 1986 a fire at a warehouse near Basle in Switzerland released some 30 tons of pesticides and the poisonous metal mercury into the Rhine River. Within a week the pollution had reached the Netherlands and the authorities shut off the drinking water supplies to cities along the route of the 40-km (25-mi) stretch of pollution. As a result of the spillage 300 km (186 mi) of the Rhine was declared dead; it will probably take ten years to restore it to life.

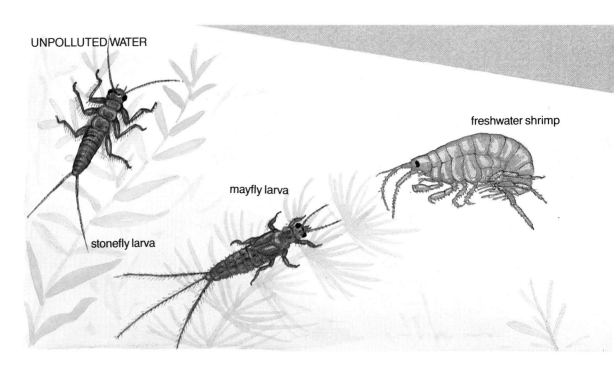

UNPOLLUTED WATER

stonefly larva

mayfly larva

freshwater shrimp

Activity: How clean is town water?

You can tell how clean a pond or river is by the kinds of animals living there. Use the net to try to collect some of them. Sweep the net quickly just under the surface and tip out the contents into a tray. After a few more sweeps examine the tray. Record what animals, if any, you find. Repeat this for different places in the stream or river, such as along the bottom and among water plants. Take samples at various places along the stream.

Remember that these animals make their homes in the mud, among the plants and under stones, so try to cause as little disturbance as possible and return all animals to the water after you have recorded them. Be very careful when working near water since it can be a dangerous place.

Some animals often found in unpolluted water include the larvae of mayflies, stoneflies and blackflies. Rather more polluted water, which holds less oxygen, will contain freshwater shrimps and water skaters. Bloodworms will be found in water with even less oxygen dissolved in it. Water that has very little oxygen in it will contain only air-breathing animals, such as rat-tailed maggots that breathe through a telescopic tube, and those that need little oxygen, like tubifex worms.

What did you see?

Did you find different kinds of animals at different places along a polluted river? If you did, can you give a reason for this? Did you find places with a lot of pollution? Where did the pollution come from? Is anything being done to clean up the waterway?

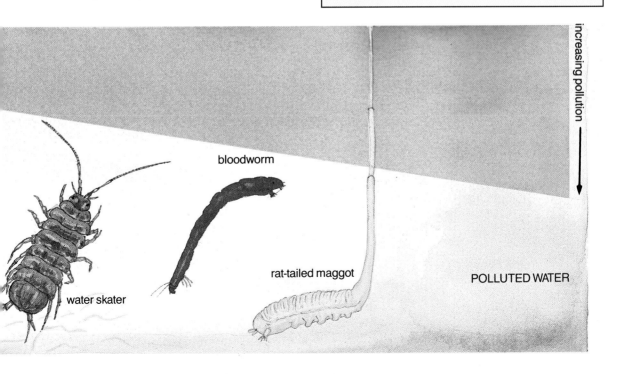

increasing pollution

bloodworm

rat-tailed maggot

water skater

POLLUTED WATER

5. Towns as habitats

Towns are excellent habitats for many forms of wildlife, and some species are now thoroughly adjusted to urban living.

Whether they live in towns or in the country plants need a supply of water, nutrients, sunlight and an anchorage for their roots. They also need shelter from strong winds. If these needs are filled, plants will grow in some surprising places. Wild plants and escapees from gardens will be found growing from cracks and crannies in walls, on roofs, in gutters and pushing up between paving stones.

Wallflowers, for instance, find enough moisture to thrive on bare cliffs of brick. The mortar itself may provide food for plant growth. Towns provide a greater range of habitats than much of the countryside. City centers contain tall buildings that provide ledges for birds to nest on, and there are many habitats within the buildings themselves. Town parks and gardens, with their open grassy areas, pond or lakes and small wooded areas, provide other habitats.

Vacant lots also provide habitats: derelict sites are quickly colonized by plants, and places where people have dumped their trash provide scavenging animals with a ready-made larder and home. Many cemeteries provide undisturbed habitats, particularly if they are overgrown. Even your own home will contain many uninvited guests that seek out warm, dry or damp parts of the house. Some even feed on the wood of the house itself or on wooden furniture.

Wherever humans go, the most adaptable animals, like rats, sparrows and cockroaches, follow. Town dumps provide a good habitat for brown rats.

Activity: Mapping town life

Draw a large map of your area or town. You can copy the map or make one from your own knowledge of the town. Mark on the map all the places where you see wildlife and distinguish one habitat from another. Parks, lakes and cemeteries are obvious places, but there may be vacant lots and garbage dumps.

Visit some of these places and record what plant life you find and how abundant it is. You may find unusual places for plants, like walls, gutters and parking lots. In each case record which plants are the most common, how much shelter the site offers and how damp the ground is. Is there much sun or is it shaded? Which plants grow well on brick and concrete and which need soil?

Do different animals occur in different places? Which are the most common in your town? Where do they live and what do they feed on?

Try to think up a food chain for each habitat. What plants are eaten by a particular insect? Is this insect eaten by, say, a spider? Might this animal be eaten by a bird? Is the final link in the chain a mammal?

Churchyards and cemeteries are usually rich in urban plants and animals. Many abandoned cemeteries have been made into urban nature preserves since they have been left to run wild. This cemetery is in London, England.

What did you see?

How abundant is wildlife in your area? What habitats are there? Are the greatest variety of insects to be found in parks and grassy open spaces? Are there any remains of the original, natural habitats, like small woodlands or marshy areas? What life was there in places with little soil and grass such as on sidewalks, under rocks and in parking lots?

6. Adapting to town life

Some creatures are naturally suited to life in urban areas. Others have had to change to fit into town life.

The peppered moth has whitish wings peppered with black dots. During the day it rests on the bark of trees and is well camouflaged. This means that its color and pattern blend in well with the bark and lichens on the tree. Lichens are very sensitive to pollution and have almost disappeared from trees in industrial areas. Without lichens, tree trunks are dark, and peppered moths resting on them are easily seen and eaten by predators.

In the middle of the nineteenth century unusual peppered moths were first noticed in polluted industrial areas. Instead of having a dotted white pattern, these moths were all black. Such unusual coloring of a particular species is quite rare, but the black version stood a better chance of surviving since it was unnoticeable in a polluted area on a dark tree. As a result, the black moths grew in numbers so that in industrial areas most of the peppered moths were black, while in the countryside the light-colored version was most common. Many moths and other insects, too, have changed their coloring as generations passed to adapt to town life, with the number of black versions increasing near centers of industry.

The peppered moth came by its name since it had black spots on a white background. At one time the light-colored form (left) was most commonly found. With the increasing pollution of industrial areas the rare black form (right) increased in numbers and became more frequently found than its light-colored counterpart.

Another example of a species adapting to town life is the house mouse. This mammal has lived with humans for thousands of years and it shows its adaptability by breeding almost anywhere if there is enough food and it is undisturbed. These creatures can evolve at a very quick rate and produce special changes for particular environments. Several cities in South America have their own particular types of house mice and they may even be found in refrigerated stores at temperatures below freezing. Such mice have developed especially thick fur to live in these arctic conditions.

The number of animals that have solved the problems of urban living is small, however,

The pigeon has adapted well to city life. It has learned that humans will feed it, and in many places it will eat from the hand. This pigeon has learned how to drink from a fountain in Siena, Italy.

compared to the great numbers of species to be found in "natural" environments. Since there is a good supply of food in towns, those species that do survive can reproduce themselves at a very fast rate. Rats are protected from the temperature and weather when living in buildings and so they breed all year round. Pigeons, even though they live outside, are protected enough so they can lay eggs at any time, even in winter.

7. Town plants

Certain plants that we think of as weeds are very successful at overcoming the problems of town life. Fireweeds, for example, are found in parks and yards and are often among the first plants to "colonize" building sites and recently created open spaces. We think of weeds as unwelcome intruders among cultivated plants, and remove them. However, they are very valuable since many town insects feed on them.

Some of the most successful colonizers of concrete and brick surfaces are lichens. Since they grow very slowly the best examples are usually found on the oldest buildings.

Damp walls provide a good environment for different species of mosses, and ferns may also do well here. Ferns establish themselves in the mortar of walls, which provides them with minerals. A variety of small animals depend in turn on the plants for their homes and food.

Many plants that grow wild in towns have escaped from gardens. Buddleia was originally introduced as a garden plant from China, but its seeds were carried out of gardens and became established on building sites and vacant lots.

Trees are a common feature of town life. Most conifers do not grow well since their needles may stay on the tree for years and their pores become clogged with grime. Deciduous trees are better suited because they lose their leaves each autumn. London plane trees have a further advantage. They regularly shed patches of bark – and with it all the soot and grime that has collected – leaving new, clean bark underneath.

Fireweed is a familiar urban plant. Each plant can produce as many as 80,000 seeds, which are carried by the wind to new ground.

Ferns establish themselves in the mortar in cracks in walls. From this they take the necessary minerals for growth.

Activity: The ecology of a wall

Select a wall to study. Old garden walls are the best but look in between the cracks of quite new walls for signs of life. Make a list of what you find and sketch the wall to show where the life is found.

Do the plants need soil in which to grow? You may find patches of lichens on older walls. These have no roots. Where do any rooted plants grow from? Are the roots visible? Examine how the plants are attached to the wall. Some may have runners, suckers or tendrils. Have they caused any damage to the wall? Think how the plants got to the wall and, if possible, collect some seeds.

Examine the other side of the wall and try to find other walls nearby that face in different directions. Does the vegetation differ? Does the material that the walls are made of affect the life they support? Compare the wall vegetation with that growing on flat stone surfaces like sidewalks.

Look to see what animals there are. What animals feed on the plants? Do some creatures make their homes in the wall?

What did you see?

How does the direction the wall faces affect the plants and animals found there? Which plants need damp, shaded conditions? Were there any flowering plants? What sorts of animals visit these?

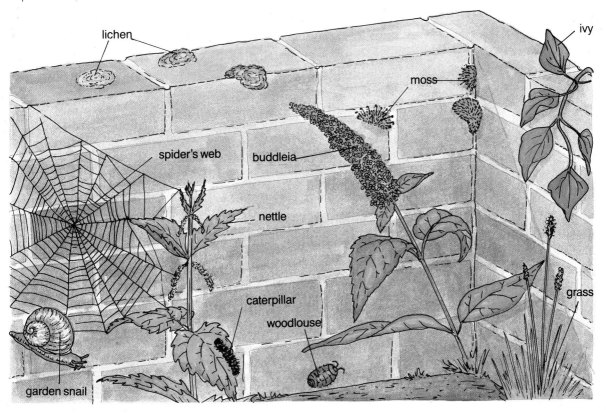

lichen · ivy · moss · spider's web · buddleia · nettle · caterpillar · woodlouse · grass · garden snail

8. Small creatures in towns

Insects and other small creatures build up to large numbers in towns. Invertebrates are creatures that have no bone inside them. Instead they have a hard, outer skeleton. Insects, spiders, slugs and wood lice are all invertebrates. Small creatures have an advantage in urban areas. They are tiny enough to live within cracks in walls and sidewalks and may flourish unseen, especially those that are active at night.

Each individual creature and species has a niche, or role to play, in its environment. The numbers of small creatures in a garden, yard or vacant lot show what that area can support and what it has to offer, like variety of vegetation and shelter. There are always plant-eaters and their predators, and those that decompose dead material. Exactly how many of each live in a particular area depends largely on the variety and numbers of plants there are.

Most millipedes feed on decaying material but some eat soft garden plants, making them unpopular with gardeners. Most of the time they are useful, bringing about and accelerating the decaying process in refuse and compost heaps.

While millipedes are herbivores, centipedes have stronger jaws and poison fangs to kill live food. Their diet includes mites, beetles, spiders, slugs and worms. Because they could quickly dry up in strong sunlight, centipedes come out only at night.

Spiders live in all types of habitats and it is not unusual to find several different species in a small garden or yard. Some spin webs to trap flies and small insects. These will be positioned where the insects are likely to be found, such as places where there is plant cover. Even bare walls will support several species. Not all spiders spin webs. The wolf spider, for instance, ventures forth at night to hunt for prey.

This garden spider has spun a web on a wall to catch small insects.

Activity: Make a pitfall trap

There are many animals that you do not see either because they are out at night or they are so quick they escape detection. With a pitfall trap you can catch these animals without harming them.

Choose a site for a trap. Possible places are in long grass, near a wall or in a vacant lot. Bury the container so that the top is level with the surface of the ground. Make sure that the container fits tightly so that there are no gaps between the ground and its sides. You can bait the trap with bread, grass or other foods. Place the stones around the rim of the container and rest the larger stone on top of them so that there is a small gap between the "lid" and the top of the container.

Set the trap at dusk and return a few hours after darkness, or wait until morning. Never use traps when the weather is likely to be wet.

Keep careful records of the creatures you find in the trap. Study them closely with a magnifying glass and count the number of legs – a part of their identification. Beetles have six legs, spiders have eight and wood lice have fourteen. Look for the differences between centipedes and millipedes. Put traps in different places and in different habitats. Try using different baits. When you have finished using a trap it is important to dig it up.

What did you see?

What variety of life did you trap? Do certain species live only in certain habitats? Do the numbers and kinds of creatures you catch change according to the weather and the time of year? Why is it important to make a hole in the bottom of the container and to dig it up when you have finished with it?

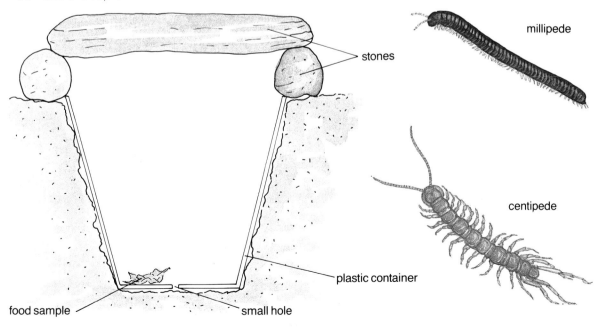

stones

millipede

centipede

plastic container

food sample

small hole

9. Flying "opportunists"

Many different species of birds have been able to use the urban environment to their advantage. These opportunists have moved in from woodland, coasts and other environments to find food and places to nest.

The sides of buildings often resemble cliff faces and provide similar opportunities for nesting. Birds that are used to cliff habitats have no problem in moving into urban areas. One of the most common city birds, the pigeon, is descended from the rock dove, which originally lived on sea cliffs, and it makes its home on ledges and in niches as it would on a cliff.

Giant flocks of starlings, numbering some tens of thousands, will swarm into cities in the autumn to roost and compete with the resident birds for the limited supply of winter food.

Swifts arrive in Northern European towns in late spring after flying from Africa. They are well adapted to making use of town life, feeding entirely on the huge numbers of insects and wind-blown spiders, which they catch in the air. On a typical flight a swift may collect 1,500 insects, packing them into a tight ball which is fed to the young in the nest. In North America, a swift that originally built its nest in hollow trees found that towns offered more ventilator shafts and chimneys than hollow trees. Today the chimney swift nests hardly anywhere else but in towns.

Pigeons and sparrows have attracted their own predators. Kestrels have found cities to be excellent for food and breeding places. From their nests on the ledges of buildings they watch for small mammals as well as small birds, which they take from gardens and bird feeders. Peregrine falcons enter cities occasionally to find pigeons.

This swallow is feeding its young in a nest built inside a shed.

Activity: The birds in a street

What you will need

Pencil, paper and clipboard.

Make a copy of the chart shown below containing the names of the most common birds you know. Leave room to add more birds to the list if you see them. Choose a quiet road or street for your study. Visit the road in the mornings or late afternoons and take the chart with you.

Quietly walk along the street looking for birds. As you see each type, put a checkmark on your chart to show its position. Repeat this study at the same time on different days to build up your results.

You might want to try another study at a different time of day or for a different habitat, such as a park or vacant lot. Compare these different results. If you keep a bird study regularly over a period of time you may be able to tell "resident" birds from "migrants."

What did you see?

Which are the most common birds in the street? What do you think attracts them to the area? Do some birds appear only at particular times of day, like dusk? Why do you think some birds might be found in one area but not in another? When do "migrants" make their first appearances? How long do they stay and when do they leave?

	road	grass	sidewalk	gutter	tree	bush	window ledge	roof	sky
pigeon									
blackbird									
sparrow									
starling									
swallow									
swift									
robin									
gull									
finch									
hawk									

A kestrel surveys the suburban scene on the lookout for small mammals and birds.

10. Wild mammals in towns

In the urban environment, plants are not the main source of food for the larger animals. Most of these have been attracted to towns by one great food supply – the waste produced by humans. Garbage cans, town dumps and litter provide the food on which whole chains of animals are based. Among the most successful consumers of this waste are rodents.

Rats and mice are thought to have come from Asia long ago and spread throughout the world with human help. The house mouse has been an urban pest for thousands of years, and living side-by-side with humans ensures it a reliable supply of food. Rats are possibly the most important mammals that scavenge. Most live in town sewers and come to the surface to feed at night. It is thought that half a mile of sewer can support at least 500 rats.

Rats carry many diseases. Since they have no major predators feeding on them in towns,

Feral (wild) cats at a city dump in Bangkok, Thailand.

diseased rodents can live for a long time, spreading their infection. In the fourteenth century fleas carried by rats were responsible for the death of one quarter of the human population of Europe.

Feeding on the occasional rodent that strays into its territory is the feral ("wild") cat. These are descended from pets but have gone wild and are well adjusted to town life. Groups of about twenty may be found living together in factories, docklands and underground garages. The Colosseum in Rome, for example, is famous for its feral cat population. In the countryside a tom cat may patrol up to 200 acres feeding on small mammals, birds and insects. In the city, where more food is available, the territory may be less than an acre. Here the cat will eat discarded food, like fish heads and potato peelings.

A mammal that has to compete with cats and share their territories is the urban fox. Throughout much of Europe, and particularly in Britain, it is a common sight and there have been reports of them in North American

towns. Realizing that humans are a wasteful species, the urban fox has moved deep into many towns, wandering among buildings and streets at night. Foxes often make their homes and raise their young in deserted buildings and in railroad embankments, using the train lines as useful "roads" into the heart of town.

The fox has learned to scavenge for meat and vegetables scrapes from humans, upturning garbage cans to remove the contents. Yet the fox will eat pigeons, sparrows, voles and squirrels, which are found in many urban parks. A recent "fox watch" in central London found over 1,300 foxes, taking advantage of the abundant food supply. Urban areas also provide a refuge for foxes since they are shot or hunted as pests in the country.

A great many smaller animals may also be found in towns. In North America skunks and chipmunks will invade backyards searching for food, and raccoons may be found scavenging in garbage cans. In the UK insect-eating hedgehogs are more easily seen in towns than in the country, frequently

discovered at night in gardens. There have also been plagues of possums in Sydney, Australia. There they steal from fruit trees and garbage cans.

The waste produced by humans has attracted many larger mammals into towns. This fox (above) is scavenging for food scraps in a trashcan in Britain. The raccoon (below) is a frequent scavenger in the United States.

11. Fly-by-night

Some of the most successful "exploiters" of towns are moths. An average garden or corner of a park may attract hundreds of different species, there being many more types of moths than butterflies. Many moths are pure town-dwellers and it seems that some are very adaptable and can change their behavior to suit food supply and competition with other animals.

Most moths are nocturnal, they search for food only at night. On warm nights hundreds of moths of dozens of species will visit flower borders in parks and gardens in search of nectar. Many will also be attracted to the windows of houses when the lights are on and will flutter around street lamps.

Bats thrive in the urban habitat, hunting for insects above gardens, parks, ponds and even sewage plants. The attics of houses provide a good alternative to their natural roosts of hollow trees and caves. On a single night a bat may eat thousands of insects on the wing, many of them pests. The bat gives out high-pitched sounds and picks up the reflected echoes from buildings and other flying creatures. In this way it can move around and find food. In colder countries they hibernate over the winter, sleeping in colonies in the attics of buildings.

The large number of rodents, small birds and insects has attracted many owls into the urban environment. They may be found hunting in parks, gardens and even in dumps and vacant lots. Although most are nocturnal, they may come out in the daytime to feed, especially in the breeding season when young owls demand a lot of food. They will nest in sheltered sites in roofs and the belfries of churches. Barn owls are reported to be breeding in the belfry at Notre Dame Cathedral in Paris.

This barn owl is leaving its nest through a hole in a roof.

Activity: Make a bat box

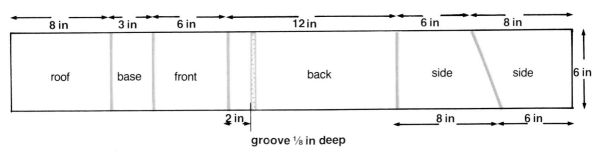

8 in	3 in	6 in	12 in	6 in	8 in
roof	base	front	back	side	side

2 in
groove ⅛ in deep

8 in   6 in

6 in

What you will need

Roughened softwood, 6 in wide, 1 in thick and at least 45 in long. A saw, hammer and nails, a tire inner-tube and thick rubber bands. You will need to take very great care when sawing and using a hammer since they can be dangerous. The help of an adult in selecting materials and positioning the box may be necessary.

Saw the wood into the shapes with the dimensions shown in the diagram. Saw a groove ⅛ in deep and 1 in wide on the back plate, 2 in from one end. This is to take the sloping roof. Saw grooves in the back and front plates about 1/16 in apart and 1/16 in or less deep. These are so the bats can walk around the inside of the box. Some grooves on the outside would also be useful.

Carefully position the sides on the back plate just below the wide groove and nail them in place. Similarly fix the front and base in position. Make sure that the slit between base and back plate is ½ to 1 in wide. This is where the bats will enter and leave the box, and a little more sawing may be necessary.

Snap the roof into position in the wide groove and secure it in place with the rubber bands. A strip of rubber tacked around the edges of the lid will give a better fit and make the box waterproof. Holes drilled at the top and bottom of the back plate will allow the box to be nailed to a tree.

Obtain permission from the owner of the tree before putting up the box. Place the box so it faces the sun during the day and there are no overcrowding branches. Large, high-flying bats prefer boxes over 16 ft from the ground. Smaller bats will use those as low as 5 ft.

What did you see?

Carefully inspect the box monthly. First, listen for any sounds. Are there the characteristic crumbly droppings? When lifting the lid be careful as bats may be hanging from it. They may also fly out of the slit if disturbed so take great care if you are up on a ladder! Identify and count the residents. Remember that in many countries bats are protected by law and it is illegal to disturb and handle them without a license. Contact a nature conservation group for advice.

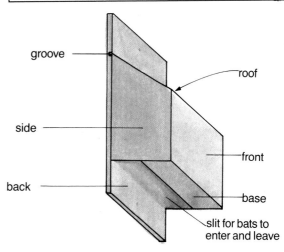

groove
roof
side
front
back
base
slit for bats to enter and leave

12. Uninvited guests

Within a human's home there are a large number of habitats and many different species of animals. Insects have been very successful in adapting to the various niches in a home, and have been living with humans for as long as there have been urban environments.

Wood-boring insects, like the furniture beetle, lay their eggs in a crack in some suitable wood, like furniture and the wooden structure of a house. When the eggs hatch, the larvae feed on the wood. After possibly years of feeding, the adults will emerge, and in severe cases the wood will crumble when touched. Termites are also pests and are often impossible to remove despite the frequent use of insecticides. The threat to house beams is great and some termites have even developed a taste for plastic, stripping electricity cables.

Another unwelcome visitor is the clothes moth. A female clothes moth lays her eggs in clothing and carpets and after a few days these hatch into tiny caterpillars. These feed on the materials, leaving holes. In the wild they probably once lived in bird's nests and animal dens and now they feed only on animal material such as woolens, feathers and fur.

Dry places attract many insects, which have followed humans from the time they lived in warm, dry caves. Silverfish are primitive, wingless insects that come out at night to scavenge for food particles on floors. During the day they live in cupboards and often behind wallpaper, which they eat to obtain the starch from the paste. However, most glues and pastes now contain chemicals that the silverfish do not like.

The housefly is the most common fly in the

Southern leaftailed geckos in a basement of a house in the suburbs of Sydney, Australia.

A house mouse with its litter of young.

home. Because it feeds and lays its eggs on garbage, manure and human food, it spreads germs. It can lay batches of 150 eggs every few weeks. The numbers of flies are kept down by spiders. Many species of spiders are found only in houses because of the dry conditions and plentiful supply of insects.

Houses, offices and factories provide a suitable home for the house mouse. It will nibble its way into the cavity between walls and beneath floorboards feeding on pantry food and even bars of stored soap. If the weather is warm enough they will live out of doors but usually migrate to the shelter of houses in autumn. They produce litters of five or six young, and in six to seven weeks these

can produce young of their own. A female can produce up to ten litters a year so it is easy to see how they can overrun urban areas.

Even less popular than mice are rats. These will also move into buildings in cold weather. Rats originated in Asia. The black rat is thought to have lived in trees, while the brown rat was a burrower. Even today, where both types infest the same building, the black rat will prefer to live in the upper stories, running among rafters, while the more common brown rat lives beneath the floor and occupies the cellar.

In tropical cities, walls and windows provide ideal territory for lizards, and there are few houses in tropical areas like the Far East that do not have geckos, snapping up insects attracted indoors by lights.

13. Parks and ponds

The parks of towns and cities are often "mini-countrysides." Most have open spaces of grass and some wooded areas and there may be mini-habitats of ponds, lakes and streams. However, most of our urban parks were not designed as nature reserves but as places where people can relax.

The urban park usually contains very few wild plants and this means that there will be fewer kinds of insects and other animals. "Butterfly gardens" are often created and ornamental flowers provide a rich source of nectar. Yet "weeds" like nettles and thistles are needed as food for caterpillars.

Parks are excellent places to see a variety of birds and sometimes squirrels. Much parkland wildlife is fairly tame and may be fed by hand. Parks are also the usual sites of the urban pond or lake. Many species of waterfowl have been introduced to these, and other species will decide to take up residence. Urban ponds are often unsuitable for waterfowl since they have steep sides instead of gently sloping ones, which prevent birds from wading onto the land.

Many country ponds have been drained or polluted, so a good urban pond can mean survival for newts, frogs and toads. Flying insects, like water beetles, dragonflies and damselflies, are also supported by the urban pond.

Stretches of water in towns can support many species of water birds. This is the pond in Kew Gardens, London, where various ducks and moorhens may be found.

Activity: Make a pond

Select a site for your pond. Make sure it is well away from trees. Dig a suitable hollow for the pond at least 6 ft by 6 ft and 3 ft deep. Be sure a "shelf" is left around the edges some 10 in deep and 10 in wide and slope the edges of the pond gently. Remove all sharp stones and sticks from the bottom of the pond.

Line the hollow with the tough plastic. Around the edges of the pond place stones and rocks. Place a clump of soil in the hollow and on the "shelf." Fill the pool with water and leave it for a week to warm up. Planting should be done in spring and summer. Water plants like pondweed should be planted in the soil, or you can put water weeds in plastic containers on the bottom of the pond.

Water lilies can be planted on the bottom and secured with some shingle. Marginal plants like bulrushes should be planted on the "shelf." Again, placing plants in containers can be a good idea, since bricks can then be used to adjust their depth.

Watch and record the colonization of the water by weeds and algae. Insects and their larvae will appear and you may find a pond snail. Introduce frogspawn from a pond that has a plentiful supply in the spring. Toads and newts may also be found breeding in the pond or hibernating in crevices among stones at the edge.

What did you see?

Why was it important to site the pond away from trees? Why is it best to stock the pond in spring and summer? What was the order of colonization? How fast did algae and weeds spread? What insects were attracted to the pond? How did a water snail get to the pond? Did water birds visit the pond? How did urban birds use it?

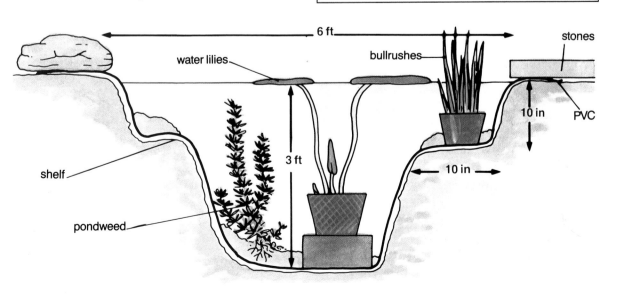

14. Vacant lots

All towns have their vacant lots, derelict areas of demolished or unused buildings and patches of land. Though not very promising places to live, the piles of rubble and bare land are soon colonized by plants. These plants find highly unlikely places to live such as blocked-up gutters, old walls and chimneys. As soon as the land is disturbed, the plants will move in. Seeds will be carried by the wind into cracks and will start growing, and those that have remained buried and inactive are brought nearer to the surface and so they start to grow.

Mosses, lichens and ferns will be among the first to colonize newly exposed land. As these plants grow, spread and die, their remains provide nutrients for later colonizers such as grasses. Fireweed, is a very successful colonizer since it can grow upon the cinders and charcoal that are left after a fire and it produces many seeds. Fireweed also produces underground shoots that allow it to form large clumps, sometimes excluding other plants.

One famous colonizer in Britain is the Oxford ragwort. This was originally brought to the country about 200 years ago from the volcanic slopes of Mount Etna in Sicily. It was kept in gardens in Oxford but the seeds escaped far beyond that city. The winds from trains on the newly created railroad system are thought to have carried it all over the country. The rubble of stones and concrete in the urban environment resemble the plant's original habitat and so it thrives in vacant lots.

In addition, nettles, wallflowers, buddleia and groundsel all thrive in vacant lots, attracting butterflies and moths for nectar and food for their caterpillars. Spiders will be attracted to hunt for insects and these will be joined by birds, which will also eat the seeds. If a vacant lot is allowed to develop further, a different type of habitat may become established. For example, apple trees may start to grow from the discarded apple cores of passersby, and bushes and trees may eventually crowd out many small weeds.

Vacant areas of land are quickly colonized by wild plants. In the foreground are Oxford ragwort, fireweed and convolvulus while buddleia is growing in the background.

Activity: Investigating a waste ground

What you will need

A notebook and pencil, together with stakes, string and a measuring tape. You should get permission to visit a derelict area before you carry out your investigation. Old building sites, town dumps and disused areas can be dangerous places and someone may own them.

Take a good look at the area and note features like walls, trees, fences, bare ground, litter and areas of short vegetation and tall vegetation. Try to find out how long the area has been derelict. If you can, compare an "old" site with a more recent one.

A good way to study the site is to make a line transect. Stretch a piece of long string in a straight line along the ground and tie it to the stake at each end. Work your way along the line to see what plants and animals you come across and measure the distance along the string where they occur. Record this information in a drawing, along with notes on the plant life you find.

Choose a line that passes over varied parts of the site. It may pass close to a wall, through an old garden, over a path, a ditch, some burned ground or a fallen tree. In the autumn, look for a variety of seeds. In the winter, find out what happens to the plants.

What did you see?

What plants grew on undisturbed soil? What plants can survive on paths? Are these of a different size and shape? Were there any areas of damp ground or water and what plants grow in and near these areas? If there was a patch of burned ground, were plants growing there? What animals did you find? How many were plant-eaters and how many fed on other animals? Did you find any plants growing on rotting vegetation? How do you think the plants got to where you found them?

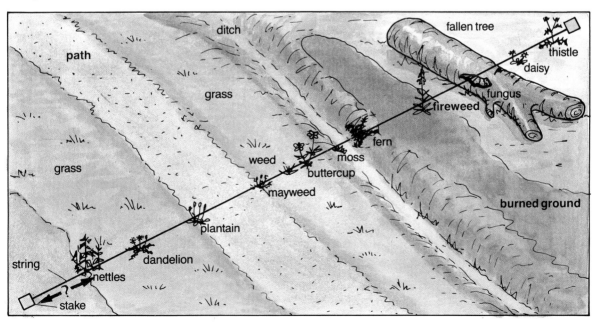

15. Encouraging urban wildlife

In many countries, the use of intensive farming techniques and pesticides has contributed to a decrease in the wildlife of rural areas. At the same time our towns and cities are providing wildlife with valuable new habitats. Animals like foxes, frogs and some butterflies may now be safer in towns than in the countryside.

In the heart of urban areas, parks, gardens, vacant lots, canals and railroad lines all provide havens for a range of animals and a huge variety of plants. As cities grow outward it is important that areas with potential for wildlife be protected.

However, few parks, gardens and roadside verges are managed for wildlife. Tidy grass and neat flowerbeds do not offer the

A hedgehog and a pet cat sharing a bowl of milk in an English suburban garden.

undisturbed variety of habitats that wildlife needs. Chemicals put down to kill ''harmful'' weeds and pests also threaten wild plants and their insect visitors.

Yet conserving urban wildlife is more than just protecting and managing special places for plants and animals. It also means the creation of new habitats, and many people are creating nature preserves in cities to benefit wildlife and make a pleasant and varied environment for humans. This could mean leaving a corner of a formal park or garden to grow wild. Larger wild habitats can also be created by planting wild flowers and plants to encourage insects, birds and mammals.

Volunteers may ''adopt'' vacant lots. Neglected areas of land may be greatly improved by efforts to encourage wildlife – like clearing away trash, cleaning up waterways and planting a variety of trees.

Activity: Make a nature garden

Often the best way to encourage wildlife is to make the most of what is already there. Choose your site, such as a vacant lot, an area of a playing field or a corner of a garden, and study what is growing. To encourage animals, think about what plants they need and why. Plant what is appropriate for the site. You should find out what the soil is like, how well it is drained, whether it is shaded and if there is any pollution. Some plants do not grow in alkaline soil, others prefer damp conditions or the shade.

Try to establish as many varieties of plants as possible, with different heights of vegetation, which will flower at different times and provide food and shelter for wildlife. Do not dig up wild plants but buy seeds packets of traditional flowers. Allow "weeds" like thistles and nettles to grow in small patches for caterpillar

food and to attract finches. Buddleia, nettles and honesty will attract butterflies and bees. Honeysuckle, primrose and night-scented stock will encourage moths to visit.

A patch of wild grasses will attract insects, including grasshoppers, and provide food for caterpillars as well as shelter for mice, shrews and voles where they can hide from the gaze of predators. A wall left to grow thick with ivy and other creepers will attract butterflies and provide a nesting site for some birds.

Old logs and tree trunks will provide a refuge for beetles. The larvae of wood-boring beetles will feed inside these and will, in turn, attract birds such as wrens. Birds should be fed regularly in the winter. Bird feeders, bird baths and nesting boxes will attract them, as will native bushes with berries and nuts. These will also encourage small mammals, as will food scraps on the ground at night.

A sheet of corrugated iron or wooden boarding laid on the ground may soon shield the nest of a vole or mouse as well as many insects, spiders and wood lice. A board leaning against a wall will provide a nesting place for chipmunks when the space between is filled with straw.

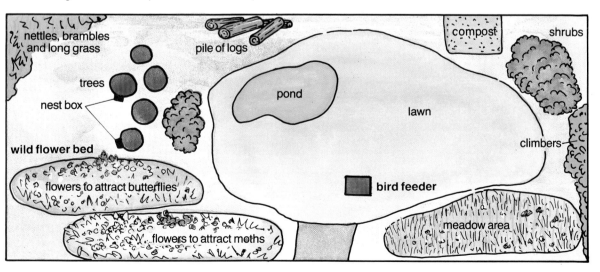

16. Trash and recycling

Even with the number of people employed to remove trash, many streets are littered with it. Discarded cans and bottles can be death traps for small mammals that climb in, looking for food, but cannot climb out up the slippery sides.

The trash we throw away may end up in dumps and trashheaps, themselves habitats for plants, insects and scavenging animals like rats and gulls. A common way of dealing with this refuse is to bury it in huge holes in the ground where it settles and some of it decomposes. However, many of the wastes are poisonous, particularly those from industries, and they can enter the groundwater and be the cause of much pollution. A solution to this has been to line such landfills with clay or plastic, but they can still leak.

The average Western household throws away about one ton of trash a year. Nearly a third of this is from plastic and paper packaging and much of the rest consists of bottles, cans and plastic containers. Newspapers and magazines – printed on paper that required huge forests to produce – are read only once and then discarded.

One way of dealing with much of this waste is to "recycle" it. This also saves valuable raw materials. Bottles, food and drink cans, paper, plastic, metal and fabric can all be reused.

Some states such as Oregon have passed laws requiring that all drink containers be returnable; in the Netherlands over half of all glass containers are recycled. Aluminum drink cans can be returned for remelting and other food cans can have the valuable tin plate removed and then be melted down to recover the steel. At present, however, only about 25 percent of the world's paper, aluminum and steel is recycled.

A mobile recycling center for glass, metals and paper in California.

Activity: A litter study

What you will need

A notebook and pencil, measuring tape and strong rubber gloves.

Make a note of everything that your household or school throws out in one day. Examine the waste and try to find out what the items are made of. Count the number of items made of each type of material and record what you find on a chart.

List those items that you think could have been used again and those that you think were unnecessary in the first place, like extra packaging. List those items such as bottles and packages that could have been reduced in number by buying in bulk. Extend your study over a week.

Find a littered area and mark out one square yard. Do not touch the litter but examine it and record what you find and the material it is made of. Repeat this for other littered sites and figure out the average amount of litter per square yard.

Find out how your local authority disposes of litter and trash, how much it costs and whether there is pollution from incineration or shortage of land to dump the wastes. Is there a place that will recycle the waste?

You could start a recycling center but you will need space to collect the items, publicity to let people know what you are doing, a method of collecting the items and a market to sell them to – perhaps a glass manufacturer or a paper factory. You may collect paper, glass (sorted into brown, green and clear) and cans (sorted into steel, aluminum and mixed metals).

What did you find?

What material is the most common among the wastes you studied? Do you think the amount of waste produced by your household is too high? What do you think the manufacturers could do to reduce the amount of waste? Are there many places near you that will recycle waste?

waste material	items	number of items	can it be used again?	was it necessary?
paper	newspapers bags cereal boxes			
glass				
wood				
metal				
plastic				
foodstuff				
other				

17. The pressures of town living

Apart from the obvious pollution of the air, water and surfaces of buildings, there are invisible pressures affecting the inhabitants of towns. Sound is a necessary part of our environment, yet too much sound at the wrong time or in the wrong place becomes "noise." When the amount of sound becomes annoying we can think of it as pollution because it can affect our health and the quality of our environment.

Noise is a major pollutant, and over the last twenty years the level of noise has risen, mainly because of increases in traffic. A survey throughout Europe looked at people's reactions to noise of various types. Among the most disliked noises around the home were the roar of motorbikes, neighbors' radios and TVs and the noise from cars and trucks.

Sound is caused by very small changes in the pressure of the air. It spreads through the air the way ripples travel out over the surface of a

Improving public transportation has been suggested as a way of reducing noise, congestion and traffic fumes.

pond after a stone has been thrown in. Noise affects different people in different ways, but at the very least it causes annoyance and at worst it can actually damage the hearing and cause intense pain.

In some places it is difficult not to make noise. Construction sites and airports are responsible for a great deal of noise pollution. Aircraft manufacturers may try to reduce the noise of engines, and airports may ban night-time flights, but this will never prevent some disturbance to people living in the area. Since noise is carried by the wind, the prevailing wind direction also needs to be considered when a new airport is planned.

The most widespread source of noise pollution is road traffic. There are some ways of reducing the noise from engines and also insulating the home against noise, such as the double-glazing of windows. However, there have been successful attempts to reduce noise by limiting the amount of traffic on the roads. For instance, many countries now ban most trucks in their towns on weekends and on public holidays.

Activity: Investigating noise

It is important to invent a scale for the level of noise you hear. Find something that makes a "normally" loud noise, like an electric bell. Stand with your ear 30 ft from the bell when it is ringing and rate the loudness with a number, say, 5. Find various other noise makers and list them in order of loudness, compared to the bell.

Make a chart like the one below and investigate the noise in your area. Visit various places and record the location on the chart. Stand still and record all the sounds you hear in one minute. Note the loudest noise. Estimate the noise value for the whole minute as compared to the bell. Visit school playgrounds, major roads, quiet streets, parks and building sites. Think of ways to reduce the amount of noise. Visit the sites at the same times of day, say, lunchtime and mid-afternoon and try to visit some in the evening.

A house in the flight path of Schiphol airport, Amsterdam.

What did you hear?

Which were the loudest and the quietest sites? What are the common sounds? Does the noise level change at different times of the day? If so, why? What sounds were louder than the bell? When do you think the level of sound becomes "pollution"?

date	time	place	sounds heard in 1 minute	loudest sound	noise value 1–10
		school playground			
		main road			
		quiet street			
		park			
		building site			
		other			

18. What can we do?

The way that a town develops and what it offers to its residents is very much in the hands of planners. These are people who design and extend towns and know the local needs for housing, offices, factories, schools, parks and roads. Most towns have a plan for their future development, and the authorities are aware of the need for a pleasant environment. In some places it is now difficult for a builder to obtain permission to cut down well-established trees to make room for new buildings. Yet in cities the competition for land to build on is intense and wild places often seem untidy and ready for development. A piece of vacant land in a town may seem a fascinating place for wildlife but it may be required for much-needed housing, particularly if the alternative is to build farther into the countryside.

Natural history and conservation groups exist in many towns, and they offer advice and even protest about development plans that are a threat to wildlife. As cities have grown outward fragments of natural habitats sometimes survive, often in the outer areas of towns where pressure to build is less intense than in the inner city. If the value of the site is known it is then often easier to campaign for its protection and designation as a nature preserve.

An urban nature preserve may be very different from the countryside equivalent, in that very rare animals and plants will hardly be seen as they are unable to stand the pressures of urban living. Yet urban nature preserves are vital since they support much of a city's wildlife and provide places where urban-dwellers may relax and enjoy the "countryside."

In this city center park in England, a meadow has been created and seeded with wild flowers.

Newly created nature gardens and ecology parks are also being developed in vacant areas and traditional parklands. Such places provide green "oases" in inner city areas, where people can learn about the natural world at first hand and where young and old town-dwellers can enjoy nature on their doorstep.

In this and other countries, interest in urban wildlife has grown rapidly over the past ten years. Some groups are directly involved in urban nature conservation, which is as much for the benefit of people as for wildlife. Such groups survey and research existing wildlife sites, campaign to save threatened areas and set up community nature parks. They also have an important role in the examination of plans for urban development.

Most countries now have laws protecting

These young people are learning about wildlife at a city center "butterfly" garden.

some green areas in towns and controlling the pollution caused by industries, traffic and noise. In the United States, planners have to submit a study of the effect on the environment of any proposed new factory. This will take into account the pollution of the air and water and the effect on the landscape, together with the possible noise pollution.

There is a limited amount that a single person can do to contribute to the quality of life in the urban habitat. Yet, you can support organizations that campaign to protect important sites and you can join a local natural history or conservation group. You could set up a group in your neighborhood to help protect and manage a local wild place and take part in practical conservation work.

Glossary

Absorb To take in a substance.

Adapt To change behavior or appearance in order to survive in a particular environment.

Bud A flower or leaf before it has fully opened.

Camouflage A natural disguise that makes something hard to see against its surroundings.

Carbon monoxide A colorless, poisonous gas that has no smell. It is given out by vehicle exhausts.

Climate The main weather conditions at an area over a long period of time.

Colonize To become established in a new environment.

Conifer A tree that bears cones instead of flowers.

Cultivated plants Plants that are grown deliberately by humans.

Deciduous tree A tree that sheds its leaves each year at the end of the growing season.

Decompose To rot as a result of the action of bacteria or fungi.

Ecology The study of how living things affect, and are affected by, their environment.

Environment The world around us, or our surroundings, including all living things. The place where an animal or plant lives may be called its environment.

Evaporation The change of a liquid into a vapor.

Evolve To change as a species over long periods of time giving rise to a new species.

Feral Having reverted to a wild state after being domesticated or cultivated by humans.

Ferns Green leafy plants that do not have flowers and grow in damp, shady places.

Fertilizer Any substance, such as manure, added to soil to increase its productivity.

Fog A cloudlike layer that forms close to the land and is made up of water droplets.

Food chain A chain of living things through which energy is passed as food.

Grain Grasses grown as food crops.

Habitat A place, having a particular environment, where plants and animals live.

Herbivores Animals that eat plants.

Hibernate To go into a deep sleep over the cold winter months.

Hydrocarbons Chemical compounds containing only carbon and hydrogen. Such compounds combine with nitrogen oxides in sunlight to give the poisonous gas ozone.

Industrial Revolution The change of Britain, Europe and the United States into industrial nations in the eighteenth and nineteenth centuries.

Insects Animals characterized by having three pairs of legs, usually wings, and a body divided into three segments.

Insulate To protect from extreme cold or heat.

Invertebrate Without a backbone.

Landfill Where wastes are buried in the ground.

Larvae (singular: larva) The grubs that develop into insects.

Lead A metal that is poisonous to life.

Lichen A type of plant consisting of an alga living within a fungus.

Mammal A warm-blodded animal characterized by having fur or hair and young that feed on their mother's milk.

Mercury A highly poisonous metal.

Migrant A creature that migrates, or changes its place of living at regular times each year.

Mosses Flowerless plants that reproduce by

spores.

Niche The position that an animal or plant holds in a community of living things.

Nitrogen oxides Poisonous gases mainly given out by vehicle exhausts. They may combine with hydrocarbons to give ozone.

Nocturnal Active at night.

Noise A sound that is loud or disturbing.

Nomadic Moving from place to place to find food or pasture.

Nutrients Minerals that are absorbed by the roots of plants for food.

Oxygen The gas that makes up nearly 21 percent of the air. It is essential for life.

Ozone A gas that is a form of oxygen. It is poisonous and is made in certain smogs in the presence of sunlight.

Pesticides Chemicals used for killing pests such as insects and rodents.

Photosynthesis The food-making process carried out by green plants. The Sun's energy is used to help convert water and carbon dioxide into food.

Pollen The powdery male sex cells of a flowering plant.

Pollution The release of substances into the air, water or land that may upset the natural balance of the environment. Such substances are called pollutants.

Predator An animal that kills others for food.

Recycle To pass a substance through a system so it can be used again.

Reproduce To bring forth a new organism of the same species; to bear young.

Rodent One of a group of animals having sharp teeth for gnawing, like rats and mice.

Roost A resting place where birds or bats sleep.

Root The part of a plant that is usually underground and absorbs water and minerals from the soil.

Smog A harmful mixture of smoke, fog and gases that may form in the air over towns.

Species A group of organisms that can breed with each other but not with other groups.

Tendrils The coiling, threadlike parts of a climbing plant.

Urban Having to do with a city or town.

Vegetation The plant life of a particular area.

Weed Any plant that grows wild and in large numbers, particularly if among cultivated plants.

Further information

Books to read

City and Suburg: Exporing an Ecosystem, by Laurence Pringle. Macmillan, 1975.

Deciding How to Live on Spaceship Earth, by Rodney F. Allen, et al. McDougal-Littel, 1973.

Ecology, by Martin J. Gutnik. Franklin Watts, 1984.

Finding Out about Conservation, by John Bentley and Bill Charlton. David and Charles, 1983.

The Future of the Environment, by Mark Lambert. Bookwright, 1986.

The Growth of Cities, by Trudy Hannier. Franklin Watts, 1985.

How Life on Earth Began, by William Jaspersohn. Franklin Watts, 1985.

Living Community: A Venture into Ecology, by S. Carl Hirsch. Viking, 1966.

Our Urban Plant, by Ellen Switzer. Atheneum, 1980.

Poisoned Land: The Problem of Hazardous Waste, by Irene Kiefe. Atheneum, 1981.

Pollution, by Geraldine Woods and Harold Woods. Franklin Watts, 1983.

Pollution: The Noise We Hear, by Claire Jones, et al. Lerner Publications, 1972.

Understanding Ecology, rev. ed. by Elizabeth T. Billington. Frederick Warne & Co., 1971.

A Walk in the Forest: The Woodlands of North America, by Albert List, Jr. and Ilka List. Crowell Junior Books, 1977.

Organizations to contact

The following organizations will provide further information including leaflets, posters and project packs. Some organize outdoor activities and there may be a local group for you to join. Remember to send a stamped addressed envelope with your inquiry.

Audubon Naturalist Society of the Central Atlantic States
8940 Jones Mill Road
Chevy Chase, Maryland 20815
301–652–9188

Children of the Green Earth
P.O. Box 200
Langley, Washington 98260
206–321–5291

Clean Water Action Project
317 Pennsylvania Avenue
Washington, D.C. 20003
202–547–1196

The Conservation Foundation
1717 Massachusetts Avenue, N.W.
Washington, D.C. 20036
202–797–4300

Environmental Action Foundation
1525 New Hampshire Avenue, N.W.
Washington, D.C. 20036
202–745–4870

Environmental Defense Fund
257 Park Avenue South, Suite 16
New York, New York 10016
212–686–4191

Greenpeace, USA
1611 Connecticut Avenue, N.W.
Washington, D.C. 20009
202–462–1177

National Audubon Society
950 Third Avenue
New York, New York 10022
212–546–9100

National Wildlife Federation
1412 16th Street, N.W.
Washington, D.C. 20036
202–797–6800

World Watch Institute
1776 Massachusetts Avenue, N.W.
Washington, D.C. 20036
202–452–1999

World Wildlife Fund
1255 23rd Street, N.W.
Washington, D.C. 20037
202–293–4800

Index

Picture acknowledgments

The author and the publishers would like to thank the following for allowing their illustrations to be reproduced in this book: Avon Wildlife Trust 40, 41; David Bowden Photographic Library 10, 18 (right); Bruce Coleman Limited *Cover* (*bottom* J. Burton, *left* J. Markham, *right* N. Tomalin), 11 (K. Taylor), 20 (J. Burton), 24 (C.B. Frith), 26 (J. Markham), 36 (J. Foote); Cecilia Fitzsimons 9, 12–13, 19, 21, 23, 27, 31, 33, 35, 37, 39; GeoScience Features 30, 38; Brian Hawkes 15, 22, 32; Oxford Scientific Films 14 (Press-Tige Pictures), 16 (P. Parks), 17 & 23 (G.I. Bernard), 25 (top/R. Redfern, bottom/B.P. Kent), 29 (Mantis Wildlife); London Wildlife Trust 18 (left); Zefa *frontispiece*, 6, 7, 8, 28, 39.